WHAT THE HECK IS OPIOID ADDICTION?

By Quincy Sirko, Clayton Russell, and Angelina Dioguardi

With Jacob Gorczyca

Dedicated to the moms and dads, sisters and brothers struggling with addiction

Text by Quincy Sirko, Angelina Dioguardi, and Clayton Russell with Jacob Gorczyca and Dr. Ellen Cavanaugh

Drawings by Lee Ann Tomlinson

Special Thanks to Janet Zellmann

Image Permissions: William Warby

ISBN 978-1-387-74194-6

Grow a Generation
Sewickley, PA 15143
www.growageneration.com

Any and all profits from the sale of this book benefit American Society of Addiction Medicine.

We all use drugs. Drugs can be helpful, but they can also be harmful if misused.

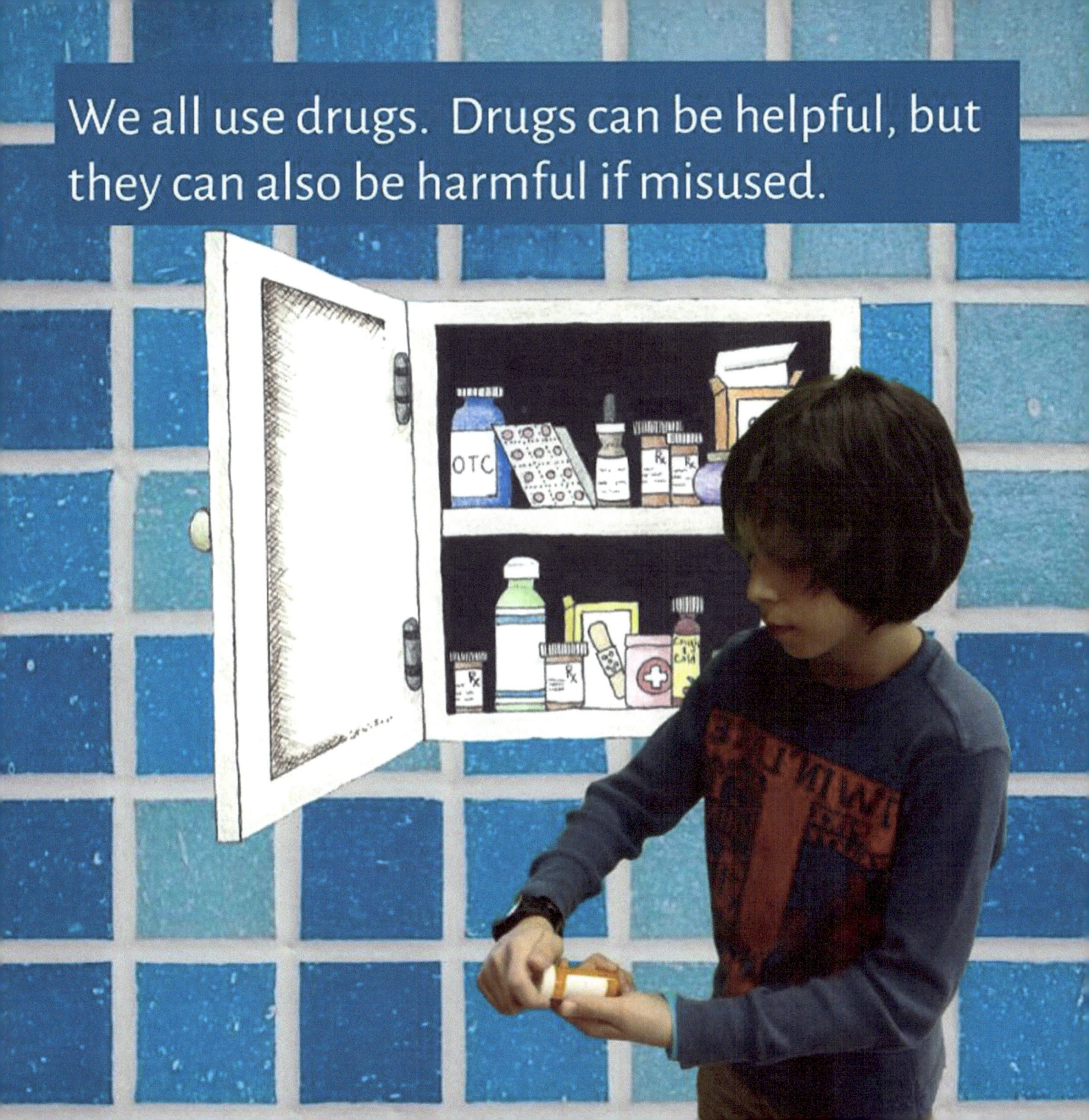

Anyone Can Become Addicted

- Some people have a genetic predisposition that makes them more likely to get addicted to anything from drugs to alcohol, video games to gambling, even cigarettes.
- Some have a chemical imbalance in their brain that interferes with their mental health.
- Some started when they were very young or when they were with friends who took drugs.

- Some started because it was prescribed. Yes, many situations of drug abuse started with prescribed pain medications!

Supporting someone who is abusing drugs is very hard.

If you know someone with an addiction, you've probably heard them promise they are quitting, only to see them slip and start using again.

Why is it so hard to stop abusing drugs? The decision to stop is the first big step. Overcoming addiction may seem impossible and often takes more time than expected.

The stakes are high! In 2016, there were more than 64,000 overdose deaths in the United States. That is almost enough people to fill Heinz Field to capacity.

Addiction is a disease of the brain. Addiction involves the compulsive seeking and use of substances despite their harmful effect. Addiction changes the structure of the brain and how it works.

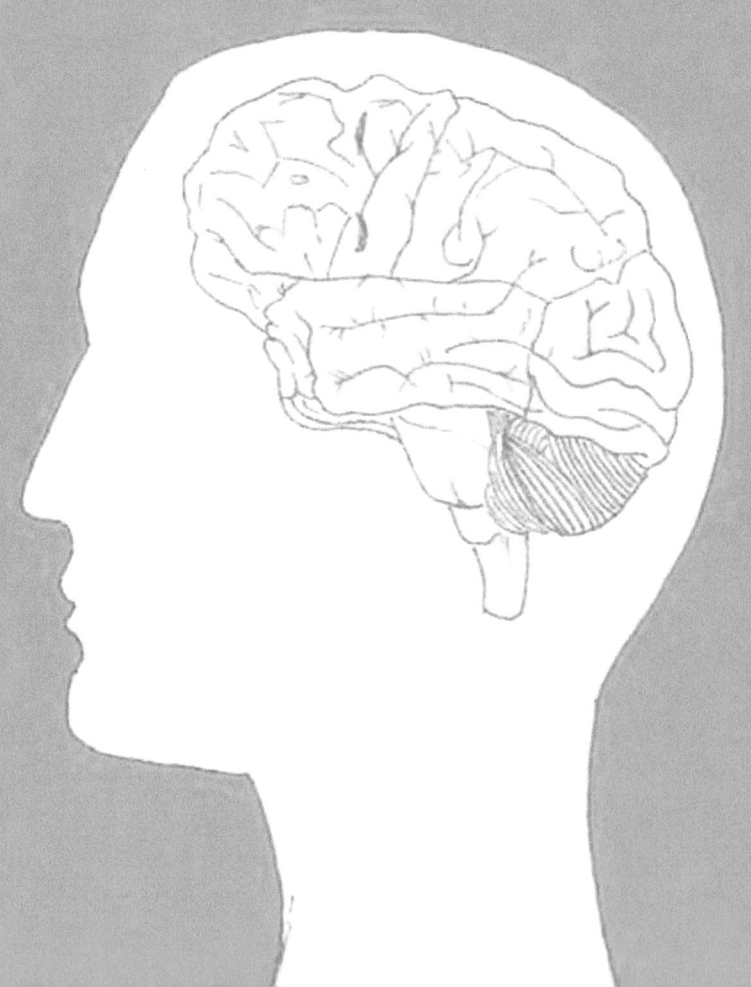

In this book, we will be focusing on opioids, as their abuse is at epidemic levels. Opioids are a group of drugs that produce relaxation, pleasure, and pain relief, but can be addictive and potentially deadly.

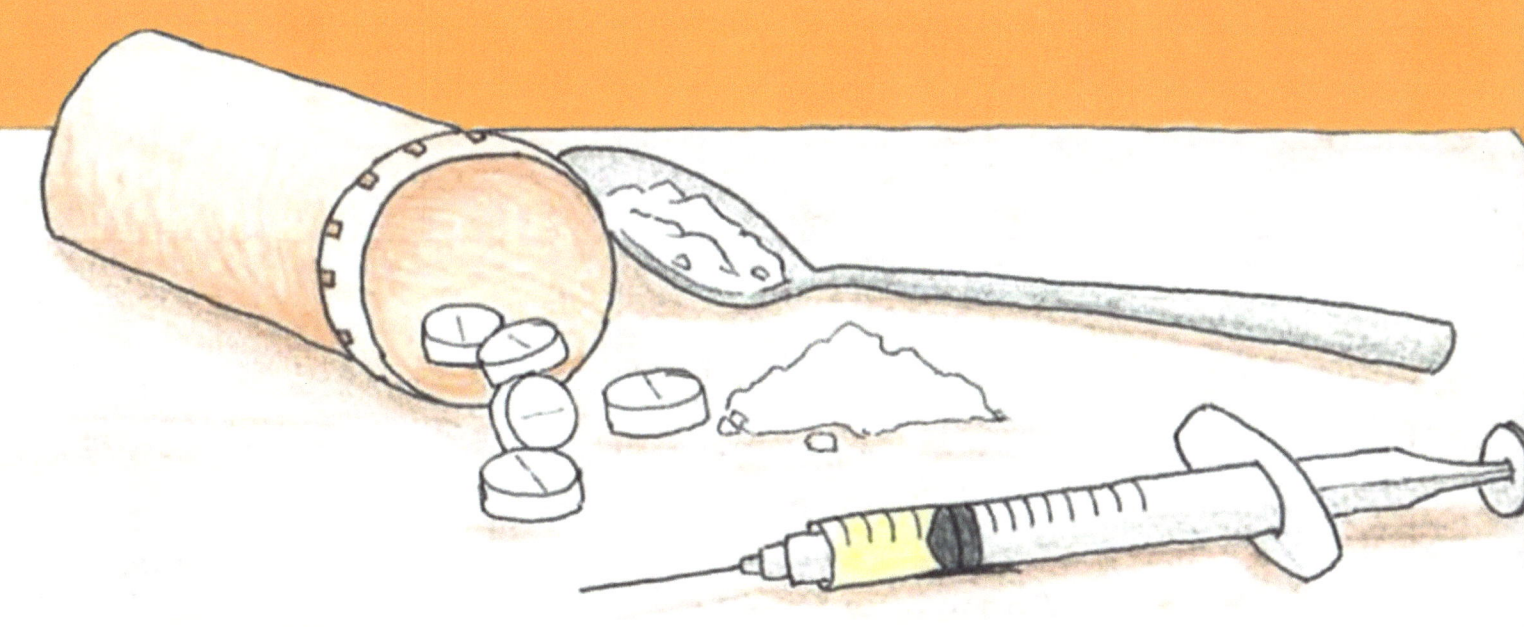

In order to understand opioids, we must understand the brain. It helps to think of the brain as a rider atop an elephant.

The rider is the ability of the brain to make decisions and take control, while the elephant represents the more basic brain functions including emotions, urges, instincts, and desires of a person.

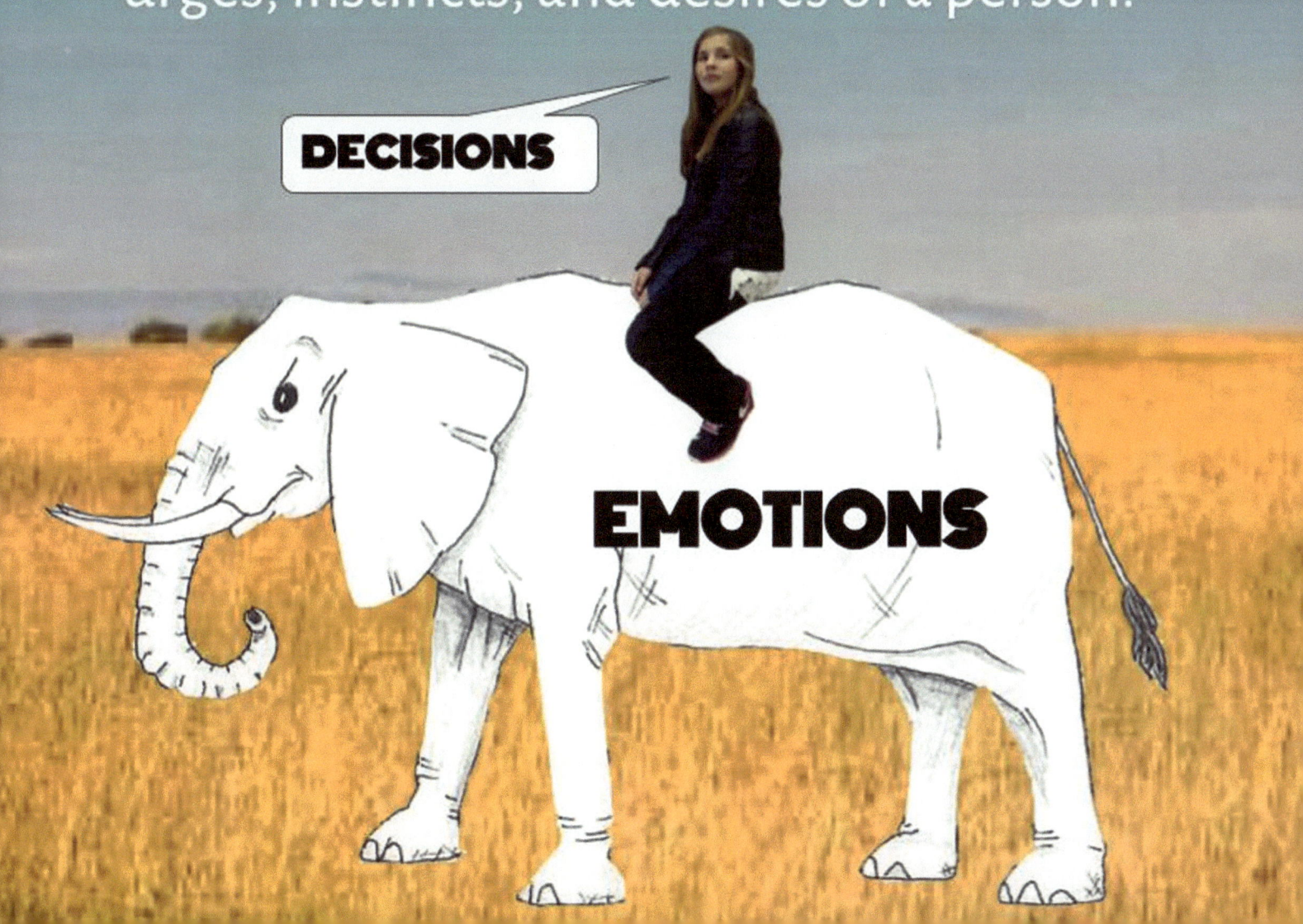

The rider has to plot the course, set goals, and motivate the elephant, which has a habit of doing what it wants.

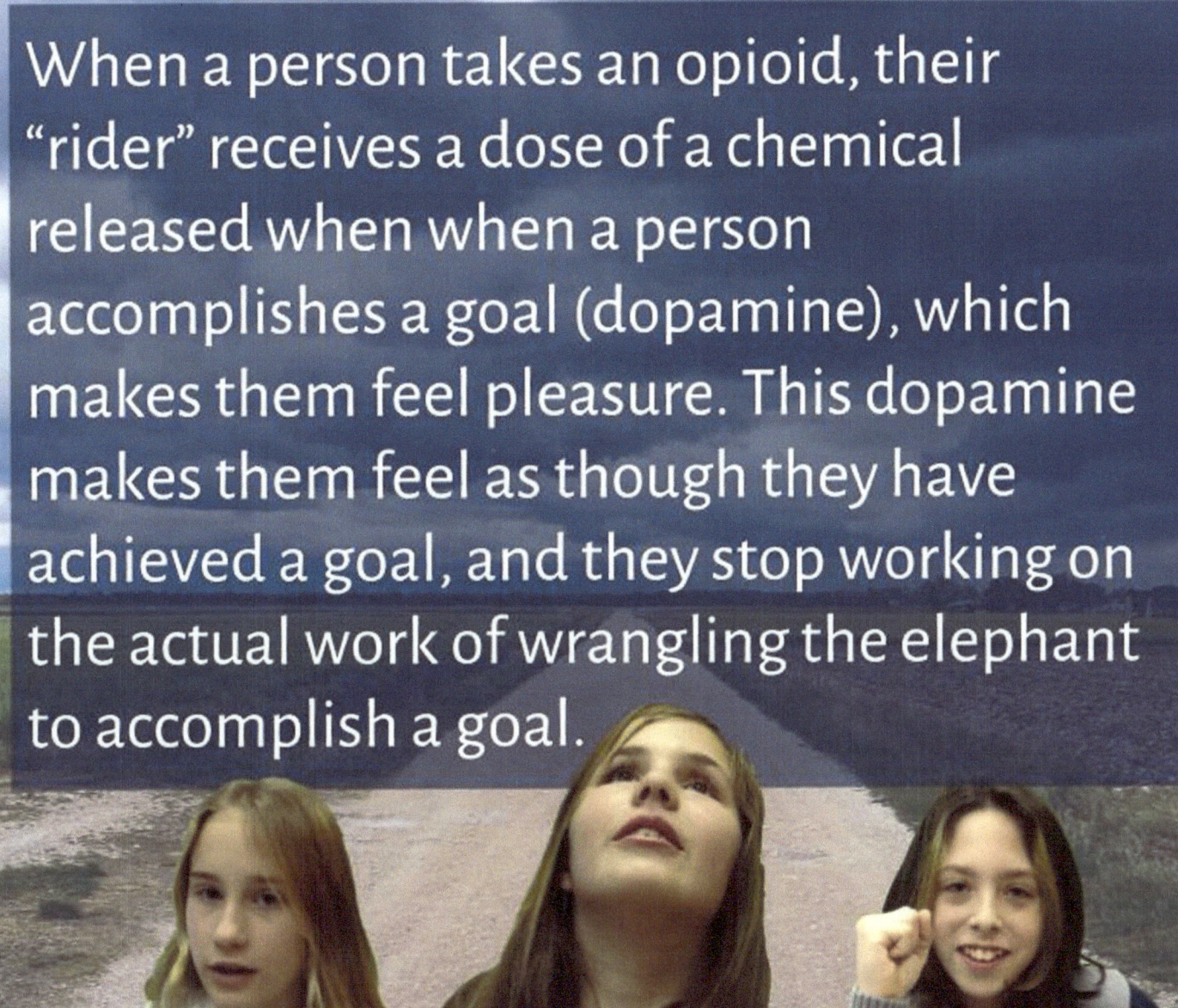

When a person takes an opioid, their "rider" receives a dose of a chemical released when when a person accomplishes a goal (dopamine), which makes them feel pleasure. This dopamine makes them feel as though they have achieved a goal, and they stop working on the actual work of wrangling the elephant to accomplish a goal.

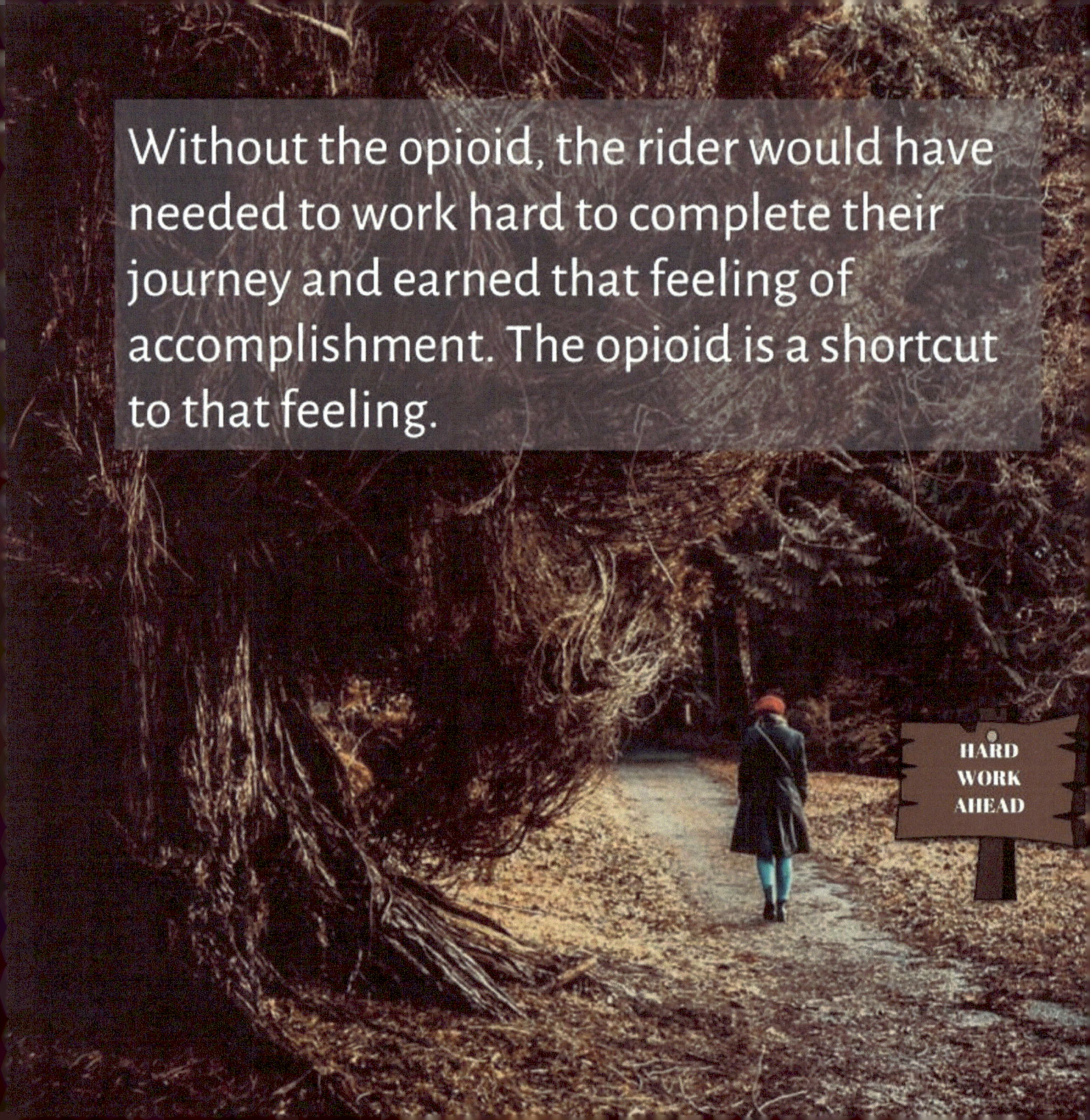

Without the opioid, the rider would have needed to work hard to complete their journey and earned that feeling of accomplishment. The opioid is a shortcut to that feeling.

HARD
WORK
AHEAD

For example, a person is invited to a birthday party for their niece. If they went, their brain would release dopamine, and they would feel the achievement of going and making someone happy. If they take an opioid, they get the same dopamine rush with no effort. They feel the pride of an accomplishment but did no work.

There are scientists who study drug abuse in a field of research known as addiction science.

They use Functional Magnetic Resonance Imaging (FMRI), scans for brains, genetic testing, chemistry, and psychology. An FMRI machine uses giant magnets rotating in a cylinder to take picture of your brain.

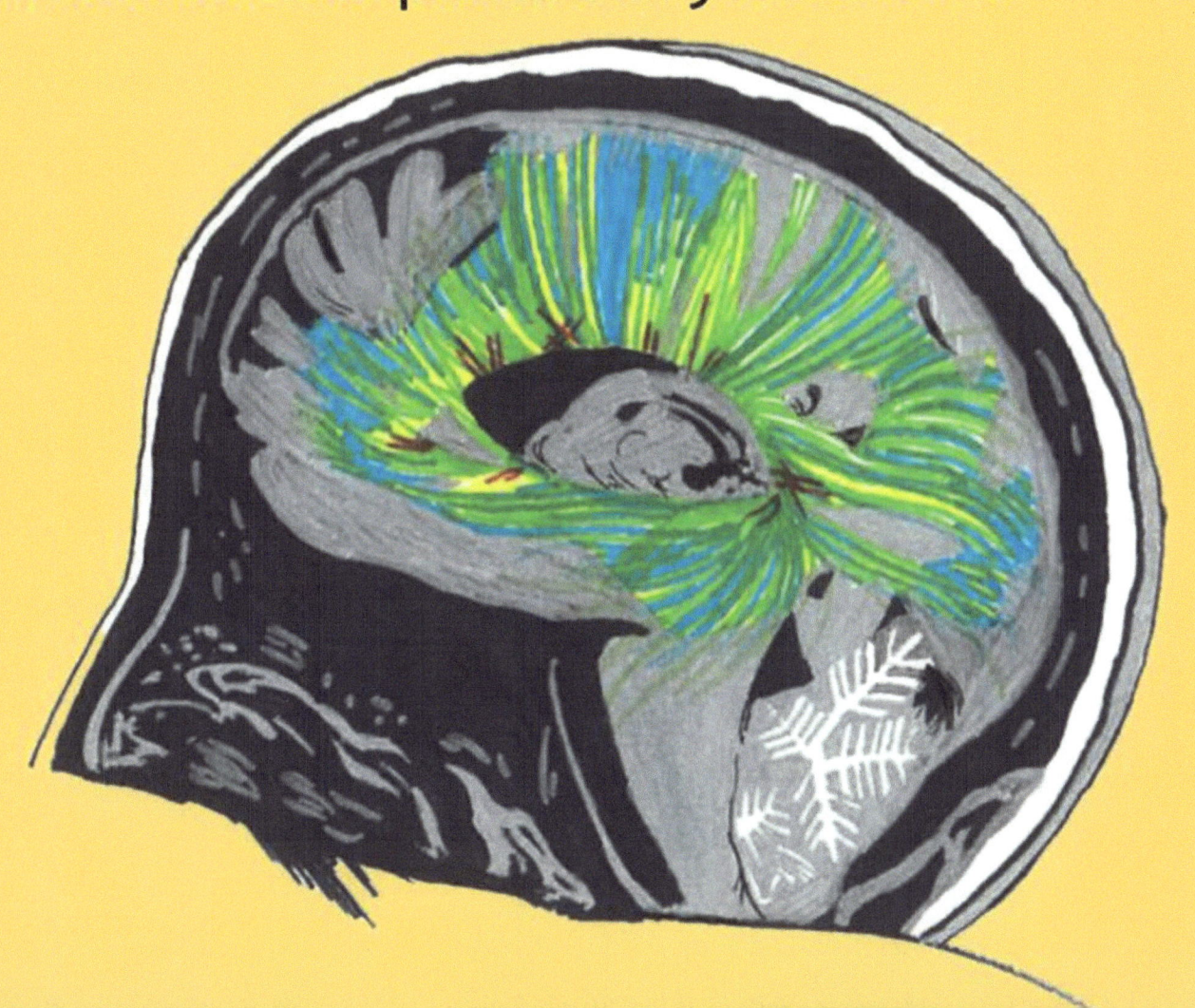

Dr. Nora D. Volkow, is the Director of the National Institute on Drug Abuse. She pioneered the use of brain imaging in identifying addiction as a disease.

When a pregnant woman takes opioids, they pass into her baby's brain. Dr. Sean Loudin, the Medical Director of the Cabell Huntington Hospital Neonatal ICU, is working with doctors and nurses to make medicines to help the babies in withdrawal.

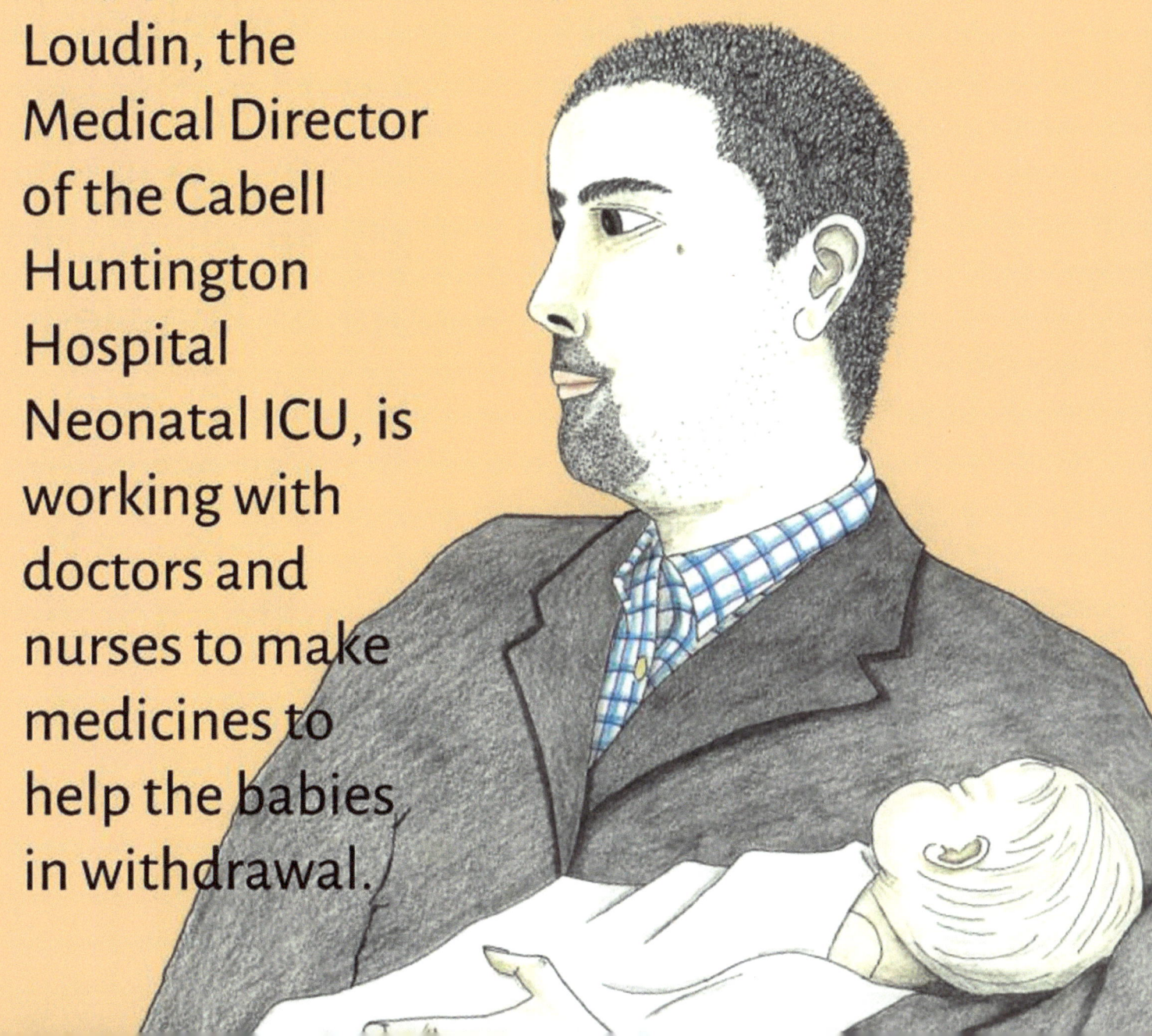

Dr. George F. Koob, is the Director of the National Institute on Alcohol Abuse and Alcoholism. He studies the roles emotions in drug abuse. His work is expanding knowledge of the reward and stress neural circuits that lead to addiction

Dr. Martha Kane is a clinical psychologist with Massachusetts General Hospital and Director of their substance disorder programs. She is rethinking how addiction programs treat young people with substance use disorders.

If you know someone who needs help, recovery is never out of reach, no matter how hopeless the situation seems. With the right treatment and support, recovery is possible.

1-800-662-HELP (4357) SAMHSA national Helpline

1-800-622-2255 National Council on Alcoholism and Drug Dependence

1-855-DRUGFREE (378-4373) Partnership for Drug-Free Kids

There are people trained as addiction specialists. The American Society of Addiction Medicine has a "Find a Physician" feature.

The National Institute on Drug Abuse (NIDA) created this brief guide containing five questions to ask when searching for a treatment program:

1. Does the program use treatments backed by scientific evidence?
2. Does the program tailor treatment to the needs of each patient?
3. Does the program adapt treatment as the patient's needs change?
4. Is the duration of treatment sufficient?
5. How do 12-step or similar recovery programs fit into drug addiction treatment?

Be wary of fentanyl. This drug is prescribed for extreme pain, but it is 50 to 100 times more powerful than other opioids. When misused, fentanyl is incredibly dangerous and often deadly.

Lethal Doses

Heroin   Fentanyl   Carfentanyl

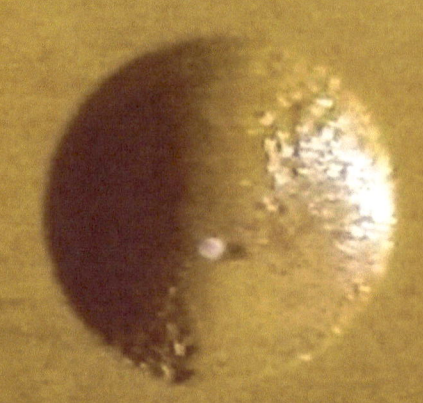

Beware of carfentanyl. Just 2 milligrams can knock out a 2,000 pound African elephant. Some street heroin is now mixed with carfentanil, making it much more deadly.

The Making of our Book

While our topic was serious, and many stories we heard were heartbreaking, we took time to have lots of fun, and tried to make this very difficult topic approachable with humor and grace.

~ Quincy, Angelina, and Clayton

Any and all profits for this book are being donated to the American Society of Addiction Medicine. ASAM is a collaboration of physicians, clinicians, and professionals dedicated to the full spectrum of addiction care. Donations go to the Ruth Fox Memorial Endowment Fund to educate physicians-in-training about the diagnosis and treatment of substance use disorders.

Baden Academy Charter School

This public charter school in Western PA works to inspire personal excellence. They cultivate the inherent gifts and talents present in all children by providing a curriculum which integrates the arts and sciences in a highly interactive, hands-on environment.

Grow a Generation

Grow a Generation partners with gifted and talented young people and teachers to make meaningful projects possible. Faculty, students, and student teams apply in their school to be accepted into the fellowship program. Once selected, they embark on a year-long odyssey to publish a book, create a digital artifact, or enter a STEM competition. Find out more at growageneration.com

www.ingramcontent.com/pod-product-compliance
Lightning Source LLC
Chambersburg PA
CBHW041302180526
45172CB00003B/939